Mu

VISITOR'S GUIDE

THE HOLY CROSS OF THE VALLEY OF THE FALLEN

José Luis Sancho

Reales Sitios de España

Palacio Real de Madrid
Belén, s/n
28071 MADRID
Tel: 91 559 74 04

N.I.P.O. 006-98-002-X

Depósito Legal: M-30618-1998
I.S.B.N. 84-7120-225-5
I.S.B.N. 84-8003-106-9

Layout: Myriam López Consalvi
Photographs: Patrimonio Nacional, Félix Lorrio
Photomechanical production: LUCAM
Translation: Nigel Williams
Printed by: Estudios Gráficos Europeos
Coordination: ALDEASA, Patrimonio Nacional

Printed in Spain

CONTENTS

PRESENTATION

Patrimonio Nacional is the institution which manages those properties of the State which are at the service of the Crown for performing representative functions under the Constitution and Laws of Spain.

The institution manages a number of Palaces, as well as several Monasteries and Convents founded by Spanish monarchs, all of great historical, artistic and cultural importance and, most significantly, of great *symbolic value*. The Royal Palaces of Madrid, El Pardo, Aranjuez, San Ildefonso and La Almudaina are used as residential and representative buildings as was intended when they were built centuries ago and it is here where His Majesty the King performs his duties as Head of State, particularly in the Royal Palace of Madrid, where this *symbolic value* is felt most strongly, as the official residence of the Crown.

In harmony with these functions, the other buildings and properties which make up *Patrimonio Nacional* have a decidedly cultural purpose and are places of study and research, as well as being open to the general public.

Both the buildings and the Spanish Royal Collections (27 in all, ranging from fans to tools and including silverware, paintings, tapestries, furniture, musical instruments, watches, etc.) are remarkable for a number of characteristics which go to make *Patrimonio Nacional* a unique cultural institution: their *particular purpose*, as they are still considered valid for representative use by the Crown; their *historical authenticity*, as they are all pieces which have been ordered, acquired or offered as gifts at some time for that particular place; their *originality*, which can be seen by the absence of replicas and imitations, and their *extraordinary artistic, historical and symbolic value*.

The combination of such impressive characteristics makes it clear to the visitor that *Patrimonio Nacional* is much more than a simple museum.

The Spanish Royal Palaces are surrounded by approximately 20,500 hectares of open land. Around 500 hectares are given over to gardens or farmland, while the remaining 20,000 hectares are forest, divided between El Pardo, La Herrería and Riofrío and part of which is open to the general public. These woodlands, mainly of the biotype *Mediterranean forest*, are of renowned ecological importance, the value of which is on a par with the monuments found in their midst.

The Royal Monasteries and Convents have been served by the same religious orders since their foundation, with the exception of San Lorenzo de El Escorial, originally of the Hieronymite Order, which was passed over to the Augustinian Order following the sale of Church lands in the 19th century. They enjoy particular importance in the history of Spain, as their origin dates back to the private patronage of the monarchs of the era.

By being open to the general public, not only do these buildings fulfil a cultural purpose, they also allow the Spanish people to comprehend their symbolic value, identify with it and consider themselves a legatee of the vast historical and artistic treasures which make up the properties of *Patrimonio Nacional.*

Collected over the centuries by the Crown, their influence in the cultural identity of Spain has been, and still is, decisive.

THE VALLEY OF THE FALLEN

For centuries, the Kings of Spain have been drawn to the Sierra del Guadarrama, the mountain range separating the provinces of Madrid and Segovia. This is why it contains buildings as evocative of the old worldwide power of the Spanish Crown as El Escorial and La Granja de San Ildefonso. It is heavy with a history that is susceptible to a variety of interpretations.

The road which leads from the Real Sitio de San Lorenzo del Escorial to Guadarrama was once the way taken by the royal retinue when, at the end of the summer "season" at the Granja de San Ildefonso, the Royal Family moved to El Escorial for the autumn "season" at the beginning of September. To the right on the way from El Escorial lies *Campillo*, then a royal possession purchased by Philip II, now in private hands. This is immediately followed, to the left, by the entrance to that lovely spot in the Madrid *sierra* known as **Cuelgamuros.** A junction at kilometre 47 on the motorway out of Madrid leads to this road and access is immediately to the right. But, in any event, the powerful form of the *Santa Cruz* or "Holy Cross" is clearly visible as a splendid feature of the landscape from a distance of some kilometres before this.

It was in Cuelgamuros where, by a Decree issued on 1st April 1940, it was decided that the Monument to the Dead in the Spanish Civil War (which had ended the year before) should be erected. It is from the deep sense of religion with which this great work was conceived by that Regime, from its situation and its object that the name comes: the **Santa Cruz del Valle de los Caídos** or **Holy Cross of the Valley of the Fallen.**

The origin, the creation and even, in many of its features, the form of this project are due to Francisco Franco, the Spanish Head of State from the time of the Spanish Civil War (1936-1939) until his death – and burial here – in 1975.

The architect who should be regarded as responsible for most of what was built was Diego Méndez, who took over as director of the project in 1950. However, it should be pointed out that Méndez was preceded by Pedro Muguruza Otaño, whose plans were followed until he was obliged to leave the project in 1949 due to ill health. For the Cross, Muguruza's plans were not used – nor indeed any of those submitted to tender in 1941 by other architects – but those of Méndez, who, moreover, made alterations to the building raised as a Monastery by Muguruza and built a new one opposite it.

With its monumentality and sheer size, the group of buildings is imposing, as is the "positive beauty" of its stonework, for, as one of the craftsmen who worked on it said, stone was then a

Campillo was purchased by Philip II from the Duke of Maqueda, lord of this small village, his idea being to use the grounds as a game preserve and the mansion as a stopping place between El Escorial and Guadarrama when he went hunting in Valsaín in the province of Segovia. In 1596 he ordered alterations to be made to the building's stately 15th-century tower, thus making for himself: "A dwelling for his Person and his servants in which in little space there was great comfort. It is square in shape and without a courtyard." (Friar José de Sigüenza). There is a fine balance between the volume of the original sturdy tower, which dates from the time of the Catholic Monarchs, and the regular arrangement of the windows, which are typical of the El Escorial area. The building is still in a good state of repair but is not open to the public as it became private property in the 19th century. From the *Way of the Cross* in the Valley there are fine views of all of this former royal possession.

View of the Holy Cross of the Valley of the Fallen.

The ***Fundación de la Santa Cruz del Valle de los Caídos*** ("Foundation of the Holy Cross of the Valley of the Fallen") was established as a trust under the Head of State and consigned to the *Patrimonio Nacional* by a Decree issued on 23rd August 1957. The 1982 law regulating the *Patrimonio Nacional* – a body in charge of managing the assets of the former *Patrimonio de la Corona* and administering the *Patronatos Reales* – has ensured that for the present the Valley of the Fallen will continue to depend on this administrative body. However, the history of the Valley, which has been studied in depth by Daniel Sueiro, is not limited to its architectural design or the special technical features inherent in its construction as mentioned by Méndez in his book *El Valle de los Caídos. Idea, proyecto y construcción* ("The Valley of the Fallen. Idea, Project and Construction"), published by the *Fundación de la Santa Cruz del Valle de los Caídos*, Madrid, 1982.

The Cross, by Méndez.

luxury – although nowadays this is much more the case. Throughout the 1950s, quarrymen from the nearby *sierra* villages worked full-time on the project. At one point, as many as three thousand men were cutting the granite whose colour betrays its place of origin: the stone from the Valley itself is brown and is considerably darker than the greyish Alpedrete granite, while that of the area closer to Segovia tends to be bluer in colour.

Work began in 1941 and ended in 1959. The official inauguration of the Cross took place on 1st April of that year, although it had been open to the public since 1st August of the previous year. Since then, together with the historic Royal Seats, the Cross has become one of the greatest and most popular tourist attractions in the area around Madrid.

The Cross and Basilica entrance from the Way of the Cross.

The "Juanelos".

The Way of the Cross shows how nature and architecture give meaning to each other in the Valley, "as suggested by the landscape, and it cannot be said to be a unit or an arbitrary creation, but rather a scattered set of architectural elements subordinated to nature..." (Cirici). The architecture is intended to endow the spot with a human – Christian – meaning, while bowing down to its grandiosity. This is perhaps why the Way of the Cross was Muguruza's favourite part of the Monument project. However, he was unable to complete it before his death, and although he left plans for all the chapels, only five – the first, second, fifth, eighth and tenth of the fourteen stations, all inspired by the Philip II architectural style – were actually built. They all offer admirable views of the Valley. Between the second and the third station there is a beautiful view of the first and of Madrid; the fourth and fifth give a good view of La Casa de Campillo; from the sixth the Cross, the esplanade in front of the Basilica and the eighth station – which crowns the crag called the *Altar Mayor* ("High Altar") and offers splendid panoramic views – are clearly visible. From this point on the steps and the road are interrupted and it becomes obvious that this beautiful project was never completed.
To continue along the route which ends at the village, the visitor will not only need two hours, a good pair of legs and to take great care in certain places (the sheer drops after the eighth station make it absolutely unadvisable to take children along) but also permission from the *Patrimonio Nacional* office in El Escorial.

The Approach Road

Cuelgamuros is one of the most beautiful and picturesque spots in the Guadarrama's spectacular granite landscape. Several rises – reaching heights of between 985 and 1,758 metres – stand out among the rugged mountains, among them the "High Altar" (surmounted with one of the Way of the Cross chapels) and, in particular, the Risco de la Nava, 1,400 metres high. This natural pyramid is christianized not only by the monumental emblem of the Holy Cross but also by the underground Basilica carved out of the rock below it.

Some of the most breath-taking views of the monument can be seen from the road which leads from the entrance to the Valley to the great flight of steps up to the esplanade spread out in front of the Basilica.

Midway between the two, announcing the solemnity of the area and serving as a reference to the splendour of 16th-century Imperial Spain, the four "Juanelos" flank the road as it runs between the pines. A metre and a half in diameter, eleven metres high and weighing fifty-four tons, these granite monoliths were cut in the Nambroca quarries (Toledo), presumably for some mechanical device which the famous Juanelo Turriano of Cremona,

The Cross and Basilica entrance from one of the stations of the Way of the Cross.

The Monastery esplanade and the Social Studies Centre from the Cross.

clockmaker to the Emperor Charles V, was to build in Toledo but never did. Instead they remained in the quarry, some unfinished until they were brought here after 1949. They were finally erected in their present position in 1953 once the idea of placing them at the entrance to the Valley or the Crypt had been rejected. To the right of the "Juanelos" begins the Way of the Cross, visitable only with prior authorization.

Further ahead, from the long *viaduct* crossing the rift, there is a splendid view of the whole amphitheatre formed by the pine-grove with the Cross on the Risco de la Nava in the centre.

Stopping on the bridge is prohibited but just after it there is a small vantage point where a car or two can park to see the Cross, the entrance to the Basilica and, dotting the other side of the gorge, the Way of the Cross chapels.

Various features of the Valley endow it with an ancestral character: its situation in the middle of a forest, the underground church, its position on a mountain, the very construction of so huge a "cruceiro" on the road that leads from the capital to Galicia – the homeland of the man for whom it was built... Even the name "Valle" given to this place dominated by the "Risco de la Nava" ("nava" in the language of the ancient Iberians meant valley) is ancestral. And this ancestral character is underscored by the romantic approach taken to this landscape that is presided over and endowed with content by an ideological symbol. The Cross blends in with the Guadarrama mountains, an area identified with the literature and nationalistic ideology of the first third of the 20th century, its granite masses being associated with the ancient basic stuff of the peninsula. It is therefore not surprising that this place was chosen when the suggestion was first made to build a national monument to those who had fallen in the Civil War of 1936-1939 (in which the Forces of the Second Spanish Republic were defeated). In all the towns and villages of Spain monuments were raised to the fallen, whose names were recorded around or at the feet of crosses erected in prominent places. This National Monument to the Fallen was raised as the greatest homage of all, although it was not until 1958 that those who had also fallen for the Republic could be buried here, on condition that they too had been Catholics.

To the right, a road leads to the *Monastery*, the Hospice and the Social Studies Centre situated behind the Risco de la Nava (see p. 43). The fork to the left, however, leads to the Basilica and the funicular.

There are a number of car parks in front of the entrance to the funicular and from there it is possible either to go up to the foot of the Cross on what is the most pleasant part of the visit (see p. 35) or to follow the road to the bottom of the great one-hundred-metre-wide steps whose two flights lead up to the esplanade in front of the Crypt or Basilica.

The Cross and the rear entrance to the Basilica with the esplanade between the Monastery and the Social Studies Centre.

The area in Cuelgamuros containing the Monument covers 1,377 hectares and was extensively reafforested after 1941. Although a serious fire in the summer of 1963 interrupted the process, reafforestation continued for many years, with the result that the trees made the Valley a noteworthy feature in the landscape. The forest is now in the care of ICONA, and, in order to avoid possible fire hazards, is closed to the public. However, its magnificent views can be enjoyed from all the roads to the Valley, the Basilica, the Monastery and the village (where the staff live) as well as from the base of the Cross.

The Cross from the esplanade leading to the Basilica.

The Exterior of the Crypt

The esplanade, an area of over thirty thousand square metres, was levelled off with part of the rock removed when the Crypt was enlarged. The great granite steps one hundred metres wide leading up to it are in two flights, ten to each flight, symbolizing the Ten Commandments.

Although Muguruza's original idea was to build a vast cross-shaped lake surrounded by graves in front of the Crypt whose waters would reflect the great monument, his final plans were not very different from those which were eventually implemented here in 1952 by the Huarte construction company under the supervision of Méndez.

However, the two architects had very different ideas regarding the façade. Muguruza, who had wished to maintain the area's picturesqueness by preserving the large rock which stood on the site of what is now its right-hand half, finally decided to build the complete semi-circle. In another of his plans, the curved, archless walls were to be articulated with tall buttresses surmounted with statues. However, it was decided to include the arches they now display and Muguruza completed the walls before he ceased to be works manager in 1949. Méndez built the semi-circle with pylons and straight wings at each end as planned but made it into a gallery. To do this he perforated the rock and made the semi-circular arches which Muguruza had originally designed as walled-in recesses, so enlivening

The Pietà, *by Juan de Ávalos, above the Basilica entrance.*

the wall with their shadows and mock frames. In this way he altered the appearance of the hill, which, instead of being a dense mass with a single entrance (into the Crypt) appeared to be tunnelled-through at the base. Before this course of action was taken, one idea had been to fill the arches with large bas-reliefs. Furthermore, the black marble background to the arches in the gallery had been designed to display in bronze letters the names of all those who had fallen and were buried there, but this was never implemented.

The Basilica entrance at the centre, built in 1950, was the first feature Méndez designed for the Valley when he joined the project as architect on the works management team in 1949. Muguruza's original idea had been a Romanesque-style frontispiece with a large number of sculptures.

From the bottom of the flight of steps – fifteen in number and sixty-three metres wide – leading up from the esplanade to the Crypt, Juan de Ávalos's large *Pietà* of black Calatorao stone (five metres high and twelve long) is clearly visible. It was Ávalos's second version of the sculpture, for when in position the first was considered too moving.

Here it is evident that although this sculpture and those at the Cross were intended to be figurative representations, their sheer volume, vigour and texture were meant to blend in with their rugged surroundings so that they might serve as a transition between the crags and the rectilinear forms of the architecture. Such was the aim of those in charge of the work, particularly the architect, who to

The bronze doors at the Basilica entrance, by Fernando Cruz Solís.

Méndez was chosen as the architect for the Valley because he was Muguruza's pupil and assistant and had collaborated with him on the work already carried out there. He had also proved himself to be an active, efficient architect and proficient where work on the Royal Seats was concerned, especially in the restoration of the Palace and the enlargement of the village in El Pardo, where Adolfo López Durán had been his assistant. But the Valley was mainly the personal creation of Francisco Franco: for his was the idea of the monument crowning the rock where the burial crypt of the fallen was to be built, his the programme for the Abbey and Social Studies Centre (the original idea to include barracks for the militia was discarded), his the choice of site, his the decisions taken on a thousand different points, and, in short, his the choice of the architects and the various projects for the Cross.

View of the nave from the screen.

convey such concepts used words like "wild", "monstrous" and "tremendous" to describe them.

Ten and a half metres high, the bronze *doors* display the fifteen mysteries of the Rosary with the twelve Apostles below and the article of the Creed traditionally attributed to each. They were made by the sculptor Fernando Cruz Solís, proposed by Ávalos and the winner of the competition of maquettes for the doors which was held in 1956.

The Crypt

In its construction, the Crypt (raised to the category of Basilica by John XXIII on 7th April 1960) gave rise to some of the greatest technical difficulties involved in the whole project. From the very beginning it was planned in two main sections: the axis (the nave or tunnel) and the large central area (or crossing and transepts). Muguruza originally intended this area to be in the shape of a cross and designed a nave eleven metres wide by eleven high with the bare rock clearly visible. It was completed in 1949, but Franco declared that the result was not as grandiose as he had expected and attributed its lack of effect to the small size of the nave where it met the crossing. Consequently, in 1950 Méndez produced a new design (on which work began in June that same year) basically dividing the nave up into various sections so that it would not give

The archangels Azrael and Gabriel, by Juan de Ávalos.

the sensation of being a narrow tunnel and increasing the size of the main one. Thus, it is in two parts, the first of which is an approach area divided up into three zones – vestibule, atrium and intermediate area – and the second of which – the actual nave – is twenty-two metres high and wide and has three chapels on each side. Another intermediate area of three sections separates the nave from the transept (with a chapel at the end of each arm), which is followed by the monks' choir at the east end. The rotunda concept was Muguruza's, while the design of the details, like that of the whole nave, was Méndez's.

The nave was made taller by removing rock not only from the ceiling but also at ground level, so that the floor lies two metres lower than the atrium. The enlargement process, which involved advancing in sections and erecting retaining walls, was extremely laborious; furthermore, calculating the stability of the perforated rock also proved very complicated as it exerted unequal pressures not only on the ceiling but also on the sides. To solve such problems Méndez abandoned the idea of the rupestrian effect sought by his predecessor and instead built the large strainer arches which, like the walls and the

The main gateway.

The Christian sense of the Monument of the Holy Cross of the Valley of the Fallen is Peace, as understood within the context of the alliance between the church and the state in Spain which lasted for many years.
This great monument was the last in a chain of grandiose churches and nationalistic military monuments which arose out of the alliance formed between the Church and the conservative middle class in Spain in the middle of the 19th century. Established by a Decree-Law of 23rd August 1957 governing the Foundation of the Holy Cross, the Social Studies Centre was the result of that same alliance.

vault, are of concrete faced with freestone. When the enlargement of the area was completed in August 1954, the total length of the Crypt was two hundred and sixty-two metres, while the nave rose to a height of twenty-two metres, the vestibule to eleven and the transept to forty-one.

Like the atrium, the *vestibule* is devoid of any adornment other than grooves and pilasters. As it serves as a reception area, it contains the gift and souvenir shop.

In recesses in the area between the atrium and the nave are two vigilant archangels, designed by Carlos Ferreira, with their heads bowed and their swords, tips embedded in the ground, held firmly in their hands, for the architect wished them to appear, "jealous of the honour of the house of God, eternally standing guard as a solemn warning to all those who enter".

On the north wall in the atrium, an inscription commemorates the Crypt's inauguration and its raising to the status of Basilica by John XXIII.

At the bottom of the eight steps descending from the atrium to the nave is the large screen designed by José Espinós Alonso and inspired by the great 16th-century Spanish works of this genre. The four pillars are filled on each side with images of saints, their colours

The Valley as a work of architecture is characterized by a use of various historical languages mixed with forms of the more recent Expressionist architectural language. The Valley is a unit only relatively, due to the fact that the grandiose aims of those in power (so often fond of architecture) took shape with the aid of a diversity of professionals and with years of delays over decisions.
As the work carried out by the architects took place in different stages, the various elements making up the whole are, to a certain extent, independent of each other. However, that the Valley was conceived as a grandiose project is evident from the immensity of its dimensions and the amount of materials used in its construction: the Crypt or underground church is 260 metres in length, while the Cross rises 150 metres above the base and 300 metres above the front esplanade.

The screen, by José Espinós Alonso, from the nave.

standing out strongly against the black iron. The screen is surmounted with military and religious emblems, while at the centre is St James on horseback flanked by angels. The inclusion of four Dominican saints in prominent positions atop the pillars suggests that Méndez wished the Monastery to be dedicated to this order.

The Saints depicted in the screen

On the side facing the vestibule, from top to bottom and left to right are:

St Mark, St Matthew, St Luke, St John, St John Chrysostom, St Vincent, St Lawrence, St Francis Xavier, St Andrew, St Cecilia, St Simon, St Francis of Assisi, St Emilianus, St Anthony Abbot,

The architect Pedro Muguruza died on 3rd February 1952, although he had been forced to abandon the project three years earlier. In 1949 he was replaced by a Management Board and because of this even the final version of the Crypt (the whole of which he had planned and by then almost completely excavated) was to be very different from his original idea. A report in 1951 on Muguruza's Crypt, which was then only eleven metres high, stated, "It lacks volume. It gives one the sensation of entering a tunnel. It is necessary to go deeper..."

"My brother planned the crypt as a part of the rock... without stone facing or anything else, but this was later seen to be impossible as small stones were constantly falling... and sometimes large ones and so the vault was finally made of stone like any other work of man-made architecture. It was later that my brother became paralysed. Diego Méndez didn't do anything at that time; my brother was fond of him. Diego Méndez is a very active man, worth his salt... and my brother had complete confidence in him. But my brother would almost certainly not have been happy with how the monument turned out. Because Méndez added some things and removed others...".
José María Muguruza, brother of the Valley's first architect, Pedro Muguruza, in an interview with Daniel Sueiro, 1974.

St Magín, St George, St Gregory, St Joan of Arc, St James and St Francis Borgia.

On the side facing the high altar, also from top to bottom and left to right are:

St Anthony, St Fructuosus, St Francis of Paola, St Dominic of the Way, St Macarius, St Edward, St Louis, St Maurice, St Ignatius Loyola, St Ferdinand, St Paul, St Augustine, St Thomas, St John of the Cross, St Teresa, St Dominic, St Hermenegild, St Peter, St Barbara and St Stephen.

Each side of the *great nave* was to have been adorned with large-scale reliefs, one of heroes, the other of martyrs, but the idea of bas-reliefs was finally discarded. Instead the Bolivian artist Reque Merubia was commissioned to design paintings on the same theme, the sketches (50 metres in length) for which are now in the General Archive at the Royal Palace. Finally, the *Apocalypse Tapestries* were hung on the walls between the strainer arches. The visitor is recommended first to visit each chapel (in the order mentioned below) and then to return up the nave to examine the tapestries more closely (see p. 22).

The nave is divided up into four sections by pairs of strainer arches. There are three chapels on either side, each dedicated to a particular advocation of the Virgin (depicted above the entrance in large bas-reliefs). From west to east, the chapels are dedicated to:

The Immaculate Conception, patron saint of the Army, by Carlos Ferreira (right).

Our Lady of Africa, by Carlos Ferreira (left).

Our Lady of Mount Carmel, patron saint of the Navy, by Carlos Ferreira (right).

Our Lady of Mercy, patron saint of captives, by Lapayese (left).

Our Lady of Loreto, by Ramón Mateu (right). She is the patron saint of the Air Force, due to the tradition that the Holy House was miraculously carried through the air from Nazareth to Loreto by the ministry of angels.

Our Lady of the Pillar, patron saint of the Hispanic world, by Mateu (left).

All the decoration in these chapels was designed by the Lapayese family. José Lapayese senior and junior designed the triptychs and frontals, which are painted on hide in the style of ancient Spanish cordovan and embossed leather decoration. The themes depicted on each of the altarpieces, once again from west to east and right to left, are as follows:

Our Lady of the Pillar *and* Our Lady of Loreto, *by Ramón Mateu.*

First on the right, the *Assumption of the Virgin*.

First on the left, the *Annunciation* (triptych) and the *Nativity* (frontal).

Second on the right, the *Marriage of the Virgin* (triptych) and the *Transit of the Virgin* (frontal).

Second on the left, the *Adoration of the Magi*.

Third on the right, the *Flight of the Holy Family into Egypt* (triptych) and the *Visitation* (frontal).

Third on the left, the *Transit of the Virgin* (triptych) and the *Presentation of Christ in the Temple* (frontal).

The twelve alabaster sculptures of the Apostles (of which there are two to each chapel) were designed by another of José Lapayese's sons – Ramón.

The Apocalypse Tapestries

The spaces between the chapels contain the eight large *Apocalypse Tapestries* (five and a half metres high by eight metres seventy wide), which are copies of those made in Brussels by Guillermo Pannemaker for Philip II. The originals, belonging to the *Patrimonio Nacional's* magnificent collection, were hung here even

St John on Patmos, *the first of the* Apocalypse Tapestries.

before the building was inaugurated in 1959, but they were affected by the damp environment and were finally replaced with copies made over a period of ten years (1966-1975) at what is now the *Industrias Artísticas Agrupadas* workshop.

Pannemaker's tapestries were thus returned to the Tapestry Museum at the Royal Palace of La Granja de San Ildefonso. The ornamental borders are also copies, except for the two nearest the crossing, which come from the Joseph, David and Solomon tapestries woven for the Royal Palace in Madrid at the Santa Bárbara Royal Tapestry Factory under the supervision of Corrado Giaquinto during the reigns of Ferdinand VI and Charles III.

The story of the *Apocalypse Tapestries* in the 16th century was a very eventful one; Philip II ordered them purchased in 1553 but on the voyage to Spain in 1559 six of the eight were lost when the ship carrying them to Laredo sank. The king ordered Pannemaker to weave new ones to replace the six that had been lost and by 1562 the complete collection was in Madrid.

The cartoons are attributed to Bernard Van Orley. Shortly before this, the prophetic Book of the Revelation – of such

The Beginning of the Last Judgement, *the second of the* Apocalypse Tapestries.

relevance to mediaeval Christian iconography – had served as a source of inspiration for a number of engravings by the German painter Albrecht Dürer.

The importance of this major series of original tapestries, the high quality of the copies and the iconographic complexity of the scenes they depict make a description of their content indispensable if they are to be properly understood.

They are discussed below in the same order as that followed by the series, which is faithful to the story of the Apocalypse.

In order to follow the story, it is necessary only to look from right to left, as the first, third, fifth and seventh tapestries are on the right, and the second, fourth, sixth and last tapestries are on the left.

1. St John on Patmos (1:9 – 5:7).

Banished by Domitian to the island of Patmos, St John the Evangelist sees the prophetic visions described in the Revelation: he is depicted on the left next to his symbol – the eagle.

In the first vision, Christ appears to John surrounded by seven candlesticks with the Gospel in his left hand, seven stars in his right, and the two-edged sword issuing from his mouth. He tells the Evangelist to take the following messages to the seven churches of Asia, symbolized here by seven angels: to Ephesus, to practise charity; to Smyrna, constancy; to Pergamos, perseverance in the faith; to Thyatira, zeal in the defence of orthodoxy; to Sardis, to combat sloth; to Philadelphia, to persevere in doing good works. The seven churches have been interpreted as the seven states of the universal church militant, and each is associated with one gift of the Holy Spirit: wisdom, strength, knowledge, counsel, understanding, piety and fear of God. Thus the prophetic message was less a reference to a specific historical situation than to the attitudes and difficulties of Christianity on earth.

In the right-hand section of the tapestry, St John sees Christ on a throne with a rainbow, the *Tetramorph* and the twenty-four elders about it. The *Tetramorph* is made up of four six-winged beings full of eyes which are the symbols of the four Evangelists: the *angel* (St Matthew), the *lion* (St Mark), the *ox* (St Luke) and the *eagle* (St John).

The Destruction of Mankind with Plagues, *the third of the* Apocalypse Tapestries.

Christ, with seven lamps of fire around him, holds before him the book with seven seals that contains the revelation of what is to come in the future.

An angel asks John, "Who is worthy to open the book and to loose the seals thereof?", and above the book John sees a Lamb with seven horns and seven eyes preparing to open it: at this point the twenty-four elders and the four beasts fall down before the Lamb singing praises.

2. The Beginning of the Last Judgement (6:1 – 7:8).

When the Lamb opens the first four seals the *four horsemen* who will devastate mankind appear. Famine, plague and war are depicted in the top left-hand corner accompanied by the *Tetramorph,* while below them is the fourth horseman – death.

When the fifth seal is opened, the martyrs, whom the angels dress in white robes at the altar of God, cry out for vengeance.

When the Lamb opens the sixth seal, the sun, the moon and the stars fall onto the earth, destroying it, and the people flee to take

The Opening of the Seventh Seal, *the fourth of the* Apocalypse Tapestries.

The Battle with the Beast, *the fifth of the* Apocalypse Tapestries.

refuge in the mountains. But the angels, led by one carrying the Cross – the symbol of redemption – hold the four winds and ask that the earth should not be destroyed until they have "sealed the servants of God in their foreheads"; the chosen, who number one hundred and forty-four thousand, are of all the tribes of the children of Israel and are depicted with an angel marking their foreheads with the sign of redemption.

3. The Destruction of Mankind with Plagues and the Adoration of the Lamb (7:9 – 10:11).

Then the Evangelist sees a great multitude of all the nations prostrated before the throne. Dressed in robes and carrying palms in their hands (centre), they cry, "Salvation to our God which sitteth upon the throne, and unto the Lamb". These are the martyrs, and they are depicted in the centre of the tapestry. St John, who is on his knees, witnesses the Apotheosis of the Lamb, whose blood is collected in a chalice.

The Triumph of the Gospel, *the sixth of the* Apocalypse Tapestries.

When the Lamb opens the seventh seal there is a great silence, after which seven angels appear with trumpets; when these are sounded the seven disasters take place in what is to be a prelude to the Last Judgement.

At top left, God the Father gives the trumpets to the angels and the effects of their sounding are depicted in the rest of the tapestry.

Hail and fire mingled with blood then begin to fall on the earth; the sun and moon are darkened; a burning mountain and a star are cast violently into the sea, giving rise to strange monsters with the body of a horse, the head of a man, the teeth of a lion and the tail of a scorpion.

The four angels of death bound in the Euphrates are then loosed on the third part of mankind, followed by an army whose horses have the tails of serpents and the heads of lions and from whose mouths fire and brimstone issue. God, through an angel that is clothed in a cloud and has feet that are like pillars of fire set firmly on the sea and the earth, gives John a book.

4. The Opening of the Seventh Seal, the Appearance of the Woman Clothed with the Sun, and the Beast (11:1 – 12:5).

The left-hand side of the tapestry depicts St John's vision of the instructions given to him by God to measure the Temple with a reed handed to him by an angel.

Also represented is the story of the prophets Enoch and Elijah, who, with their prophecies by now fulfilled, are killed by the Beast and immediately ascend to heaven (at the centre of the tapestry) filling the wicked with dread.

The Glory is depicted at the point where the seventh angel sounds his trumpet and the elders prostrate themselves crying, "The kingdoms of this world are become the kingdoms of our Lord, and of his Christ, and he shall reign for ever and ever".

Then the temple of God is opened in heaven, the Ark of God's testament is revealed, and a woman appears who is "travailing in birth and pained to be delivered". She is clothed with the sun, the

The Marriage of the Lamb, *the seventh of the* Apocalypse Tapestries.

moon is under her feet and she wears a crown of twelve stars upon her head.

A dragon with ten horns and seven heads, each with a crown, attempts to devour the child as it is born but the infant is taken up by an angel to God's throne to rule all nations for ever and ever with a rod of iron.

In the background, Babylon is destroyed.

5. The Battle with the Beast (12:7 – 14:5).

The woman clothed with the sun manages to escape thanks to the eagle's wings given to her by St Michael; this scene appears at the centre of the tapestry, while, on the left, the battle between the Beast with its demons and Michael with his angels is depicted.

The Beast is vanquished but continues to pursue the woman. In the background, another beast similar to the previous one, with a number of heads and the name of blasphemy upon them, emerges from the sea. Another beast emerging from the earth causes men to worship the first.

These beasts are similar to leopards, but have the skin of a bear and the mouth of a lion. The first is the spirit of Evil, which gives its power and attributes to the others; the second, symbolizes apostasy and idolatry and has a great wound which seems healed on one of its heads; and the third represents Rome and the emperor Diocletian in particular, whose oppressive domination is depicted in the foreground on the right.

John then sees the Lamb on Mount Sion and the 144,000 redeemed worshipping it and singing.

Christ in majesty, the *Tetramorph* and the twenty-four elders are the central theme of all the scenes.

6. The Triumph of the Gospel (14:6 – 16:10).

The angel, at the top left of the tapestry, reveals the Gospel to men, filling them with the fear of God, while in the centre the wicked are tormented with fire and brimstone in the presence of angels and the Lamb.

On high, Christ with a sickle in his right hand makes ready the Last Judgement, all of which is symbolized in the allusions to the harvest and the vine as a time for reaping and gathering the grapes of the earth by the Lord.

Depicted on a cloud, one of the figures of the *Tetramorph* – in this case the lion, symbolizing St Mark – orders the angels to release the last seven plagues on earth, the sea and the rivers and fountains of water.

The Triumph of the Church over the Demon, *the eighth of the* Apocalypse Tapestries.

7. The Marriage of the Lamb (16:10 – 19:18).

The angels pour out the last three vials of the plagues first on the air, then on the Euphrates and finally on the throne of the Beast, whose kingdom is filled with darkness. After this Babylon is destroyed. The three beasts which are referred to in the fifth tapestry then vomit unclean spirits into the Euphrates, on whose bank sits the figure of Babylon, the great harlot, offering the monarchs of the earth, who lie intoxicated at her feet, the cup of abominations, which is filled with blood.

In the background at centre, the great harlot is enveloped in flames while the angel casts a millstone into the sea.

Above the harlot, Christ in majesty, sitting in glory surrounded by the Tetramorph and the twenty-four elders, celebrates the destruction of Babylon.

Below this is the supper celebrating the marriage of the Lamb to the Church.

The armies of Christ follow the Saviour, who is enthroned on the fountain of the water of life and from whose mouth comes the sword of the Word of God. Below, Babylon, seated on the Beast, raises the cup of abominations.

8. The Triumph of the Church over the Demon in Paradise (19:19 – 22:21).

The armies of Christ, on white horses, make war on the seven-headed Beast, who symbolizes the forces of hell; the Beast is defeated and an angel chains it. Another angel shows the Evangelist a turreted fortress, the symbol of the victorious Church, with the Apotheosis of Christ above it.

This colossal work was carried out in very difficult years, due to the state of the country after the Civil War, the effects of World War II and Spain's subsequent isolation. In spite of everything work advanced, with the various parts of the project being awarded to different companies by tender. For labour the contractors employed paid prisoners-of-war who had opted for the "remission through work" system, according to which they could work off up to six days of their sentence with one of toil, although in general the ratio was three for two.

The final section of the nave; in the background, the crossing and choir.

The crossing with the High Altar.

The Eternal Father is seen sheltering holy Jerusalem under his mantle, while below, the damned are separated from the chosen by angels at the gates of Hell.

The last section of the nave before the transept.

Situated on either side of the last section of the nave before the transept, four large figures dressed like the mourners found on mediaeval tombs forewarn the visitor that he is approaching the mausoleum at the crossing.

These figures, in which Luis Antonio Sanguino and Antonio Martín achieved a sharp contrast between the rough texture of the garments and the polished anatomical forms, represent the Army, the Navy and the Air Force and their respective militias.

The Transept

The construction of the transept was awarded to a different company from that working on the nave; thus both areas were made separately, each company having its own way of access to the area – one being the tunnel to the rear esplanade now used by the monks when they enter the Crypt from the Monastery.

With the same freedom as before, Méndez made changes to Muguruza's project for the transept, giving it an Herreran-style air by means of large Doric pilasters (although with features, such as the large splayed arches and sharply projecting voussoirs, devoid of historicism). The proportions of the transept diminish the objectively large dimensions.

Estimating the reaction of the hollowed-out rock, not only in this great cavity, but also in the chapels, tombs and connecting passages, and calculating the concrete reinforcement necessary to control it, gave rise to a great many technical problems.

For the dome, Méndez rightly decided to line the great hemispherical cavity with a shell 33 metres in diameter covered on the outside with asphalt fabric and positioned at a distance of two metres from the concrete-covered rock so that any moisture would be channelled into a conduit and ensuring that the mosaic adorning the inner face of the dome would remain dry. This method was not used for the nave as the original intention had been to leave the rock visible in order to stress its character of a Crypt deep in the heart of a mountain. As the whole vault and above all the strainer arches are covered in concrete, achieving the same effect was hardly possible; consequently stone slabs were placed between the freestone-faced concrete arches to simulate such an effect.

The archangel Raphael, by Juan de Ávalos.

Christ in Majesty with the Virgin Mary and Saints, *the mosaic on the dome above the crossing, by Santiago Padrós.*

Méndez was thus to see his plans for the architectural purity of the walls thwarted. The large recesses between the pilasters were filled with four great bronze statues by Juan de Ávalos; eight metres high, they represent four of the archangels. The three which look heavenward – *Michael, Gabriel* and *Raphael* – are generally well-known, while it is necessary to recall the symbolism of the fourth archangel's name – *Azrael,* he who leads the souls of the dead before God – to understand why his head is veiled and bowed (in an attitude similar to that of Ávalos's first, rejected, model of the *Pietà*). Perhaps here more than in any other of Ávalos's work in the Valley, signs of his *Art Deco* training in the 1930s are most evident.

The *mosaic* adorning the dome is an extraordinary work of art, not only due to its artistic excellence but also to its size. It was made by Santiago Padrós, who took four years to position the more than five million tesserae.

Christ in majesty, surrounded by angels and various groups of saints and martyrs, is the central figure. To his right, the large figure of the Apostle St James leads a procession of Spanish saints, including SS. Isidore, Domingo de Silos, Dominic, Raymond of Peñafort, Ignatius Loyola and Teresa of Jesus, while to his left St Paul heads a large group of martyrs, many of whom are also Spanish.

On both sides, above the side arches over the crossing, are two other groups of figures; although anonymous they are the fallen. The group on the right consists of heroes (with standards

The archangel Michael, by Juan de Ávalos.

The Immaculate Conception, *a detail from the mosaic on the dome above the crossing, by Santiago Padrós.*

Christ in Majesty, *a detail from the mosaic on the dome above the crossing, by Santiago Padrós.*

and other details at the bottom alluding to the war in which they fell) while that on the left is made up of religious and lay martyrs. Facing Christ, the Virgin Mary leads other groups towards God.

Below the dome, the *High Altar* is dominated by a large *Christ on the Cross* sculpted by Beobide and partially painted by his master, Ignacio Zuloaga. It was placed in position on 16th April 1957. The juniper for the cross was chosen personally by Franco in the mountains of Segovia and carved in his presence. The *altar* displays gilded bas-reliefs of the *Entombment* and the *Last Supper* which were carved by Espinós.

At the foot of the altar on the west side is the tomb of José Antonio Primo de Rivera, Marqués de Estella and founder of the Spanish Falange party, whose ideology united the forces which were ultimately to overthrow the Second Spanish Republic. His body was moved to this resting place from the Basilica of the Monastery of El Escorial on 29th March 1959, three days before the Crypt was officially inaugurated.

(Left) the Choir.

(Right) the Chapel of the Entombment.

On the other side of the altar and in front of the choir is the tomb of General Francisco Franco Bahamonde, Head of State until his death on 20th November 1975. The tomb was completed as early as 1959.

At the ends of the transept are the Chapels of the *Blessed Sacrament* (left) and the *Entombment* (right). In the latter, which is the last station on the *Way of the Cross* through the Valley, is an alabaster *Recumbent Christ* flanked by *Our Lady* and *St John*, all by the Lapayeses.

The choir

Forming a semi-circle, the monks' stalls consist of seventy seats set out on three levels.

Both the architectural adornment and figures in the choir are classicistic and were carved by the Lapayeses in walnut and lemon wood. The upper relief depicts scenes of the Crusades. The backs of the two seats of honour display alabaster images of St Benedict (abbot's chair) and St Francis.

The Cross

The base of the Cross is reached by the funicular, which was not installed until several years after the inauguration of the monument.

The funicular
The construction of a funicular to the base of the Cross for use by the large number of visitors as an alternative means of access was considered in 1975. Diego Méndez criticized such a break with the original idea for the Risco in his book *El Valle de los Caídos. Idea, proyecto y construcción* ("The Valley of the Fallen. Idea, Project and Construction"). However, the funicular was made to blend in harmoniously with the monument. The work was directed by Ramón Andrada.

The Cross, illuminated at nightfall.

The competition for the preliminary designs for the Cross was judged on 22nd February 1943 by a panel which included Francisco Íñiguez Almech and Luis Gutiérrez Soto.
"The projects submitted by Del Valle, Fernández de Heredia, Romero, Pericas, Robles and Redondo were disqualified for not complying with the conditions of entry. Those submitted by Fernández Shaw, García de Alcañiz, Pigrau, Ricart, Illanes, Arrate, Blanch, García Ochoa, Cárdenas Valentín and Casulleras were considered unsuitable, while those of Olasagasti and of Francisco and Manuel Prieto Moreno, although well-planned and even beautiful, were judged as having the wrong approach. Those of Barroso, Muñoz Monasterio and Herrero Palacios, Moya, Huidobro and Thomas, Feduchi, García Lomas, Roa and Quijano, Corro, Faci, and Bellosillo were regarded as satisfactory.
Singled out were those of Moya and his team, who won first prize, and Corro and his team, who won second prize".
Another very noteworthy project was that by Francisco Cabrero, who had just returned from Italy, where he had been studying. Muguruza did not allow him to submit it, however, as he was still not a qualified architect.

A great technical achievement, the 40-metre span of the arms of the Cross was the part of the project which involved not only the largest number of plans but also the longest delays over decisions. In a national competition for designs for the Cross which was held in 1941 twenty projects were submitted, some being extremely noteworthy. However, neither Muguruza nor Méndez actually took part in this competition.

The competition was won by Luis Moya, Enrique Huidobro and Manuel Thomas, but due to further misgivings regarding the project, Muguruza was asked to seek another solution. Before he was removed from the Valley project, Muguruza produced at least two preliminary projects (the maquettes for which still exist) but neither was accepted.

Although Méndez, Mesa and Prieto Moreno formed the new Board in 1949, they could not reach agreement on a joint design, and each was finally asked to submit his own plans. On 6th January 1951 Méndez's project was accepted.

The construction of the Cross, which cost 33,661,297 pesetas and was undertaken by the Huarte company, began in 1950 and ended in 1956. The structural problems mainly involved: the base, which is firmly anchored to a solid block of reinforced concrete; the shaft,

(Left) the Virtues Fortitude and Temperance, *by Juan de Ávalos, at the base of the Cross.*

(Right) the Virtues Prudence and Justice, *by Juan de Ávalos, at the base of the Cross.*

which was made without scaffolding and reinforced from the inside as construction progressed; and the assembly of the arms, consisting of a metal framework based on triangles which was first tested out at ground level.

During this process the engineers Ignacio Vivanco and Carlos Fernández Casado, whose calculations for the structure proved to be indispensable, acted as consultants to Méndez.

Ávalos's sculptures – admirable due to their extraordinary scale – were essential to Méndez's final version.

There are four figures at the corners of the base of the Cross and another four, at a higher level, at the bottom of the shaft. The former represent the *Four Evangelists* while the latter depict the *Four Cardinal Virtues*.

All were made according to a traditional system known as "pointing" used by sculptors to make copies or enlargements of a maquette on a gigantic scale.

Once the small models had been approved, Ávalos made others on a much larger scale, from these making the drawings which enabled him, through a system of coordinates, to plot the position of each stone. The stones were laid in horizontal courses which in turn were fixed to a concrete frame.

Ávalos later praised the skill and good workmanship of a number of his assistants in implementing his designs through this very complicated process.

Each of the *Four Evangelists* is depicted with his symbol – *Matthew* with the angel, *Mark* with the lion, *Luke* with the ox and *John* with

When Méndez designed the Cross, he used some of his competitors' ideas. For example, from the project which won the second prize in the competition (by Corro, Faci and Bellosillo) he took the figures at the base, stylizing forms for the Cross, however, by giving the shaft and arms a Greek-cross section.

St Mark, by Juan de Ávalos.

The total cost of the work according to Diego Méndez's figures:

1. Crypt	356,171,933
2. Cross	114,704,043
3. Façade	111,851,898
4. New Monastery	89,895,698
5. Various expenses and additional works	82,326,356
6. Monastery	60,330,159
7. Arcade and rear esplanade	59,664,526
8. Esplanade	46,774,744
9. Roads	30,330,431
10. Staff housing	20,867,945
11. Way of the Cross	19,510,929
12. Stone cutting and carving	16,136,886
13. Rear door of the Crypt	14,425,009
14. Fees (architects, civil engineers and public works assistants)	11,242,954
15. Rear access gallery to Crypt	9,698,364
16. Dam	7,686,131
17. Conduits	6,138,775
18. Drains and sewers	6,104,648
19. Crypt entrance	5,763,547
20. Crypt entrance door	3,662,240

the eagle. Originally the sculptor wished to depict St John – the sculpture which Méndez liked best – with a long beard, but finally did not do so.

The difference of opinion between the sculptor and the architect regarding the degree of refinement as far as the placing of the courses of stones was concerned is evident in a comparison between the figure *St Mark* – very much to Ávalos's taste as the stones were placed with great care and the joins are hardly noticeable – and that of *St Luke*, in which the contractor's haste to complete the work contributed (much to the sculptor's chagrin) to the final "rustic" effect which the architect desired but Ávalos did not find at all convincing or to his taste.

The *Four Cardinal Virtues* which stand at the base of the shaft – *Prudence, Justice, Fortitude* and *Temperance* – were depicted by Ávalos as male figures.

21. Arcade irrigation system	2,464,039
22. Woodland conservation	2,158,788
23. Fencing	1,546,276
24. Steps from road to esplanade	1,535,304
25. Entrance gate	1,269,339
26. Technical office	861,245
27. "Juanelos"	860,094
28. Expropriations	653,483
29. Workers' accommodation in villages	620,000
30. Tapestry competitions	480,357
31. Reliefs competitions	260,000
32. Maquettes	257,435
33. Maquette competitions	132,915
34. Foundation's shield	72,500
35. Reforestation	1,325
Total pesetas	1,086,460,331

But as this amount is the result of funds spent over a period of almost twenty years with the consequent variation in the value of the peseta, it has been estimated as being equivalent to 5,500 million pesetas (1975 value). One part of this total that is extremely difficulty to calculate corresponds to donations made during the Civil War.

St John the Evangelist, by Juan de Ávalos.

The initial budget assigned to Ávalos was just over eight million pesetas, "And then Diego Méndez said me: 'Look, if you like we'll add nine hundred thousand pesetas to make nine million'. So I signed the contract for the Valley's nine gigantic statues for a fee of nine hundred thousand pesetas. This was without getting anything for supervising the placing of the stones; as the size of the statues was so big, this had to be done by coordinates. In horizontal courses. I'd solved that problem by reducing each sculpture to coordinates so that they could be positioned easily even by people who knew absolutely nothing about sculpture; all they had to do was keep to the measurements... And I had to find a workshop suitable for modelling and enlarging where the figures could be made to their final size. Méndez had said to me: 'Now don't worry, make the sculptures a tenth of their final size, then they can be enlarged, and it doesn't matter if they have a rustic look about them'. All this talk about 'rustic' effects in art, I don't believe in it. There's the 'rupestrian' effect, of course, but this is the 20th century and I don't like rustic effects... I had to teach people what to do, but I made my own sketches and enlargements, many of these according to their final size. I've got photographs of the heads, which were modelled at four metres fifty. I modelled the *Pietà* at its final size and used 35 tons of clay". Juan de Ávalos, in an interview with Daniel SUEIRO, *La Construcción del Valle de los Caídos*. Sedmay, Madrid, 1976.

The Cross. In the foreground, St Luke, by Juan de Ávalos.

The Cross. In the foreground, St Matthew, by Juan de Ávalos.

Initially Ávalos estimated the cost of the eight sculptures at nine million pesetas but the incidental expenses involved in so ambitious a project led to a final cost (for which the sculptor was assigned additional funds) of around seventeen million pesetas. However, even taking current prices into account this was ultimately not an excessively high amount.

A spiral staircase and a lift inside the shaft give access to the top of the cross but are too small for use by the public. However, it is not necessary for the visitor to actually go up to the top of the cross to see that the view from the base of the Cross of the *Sierra* in the distance, the plain running down to Madrid, and the nearby Cuelgamuros pine groves is truly magnificent.

Also clearly visible from here are the buildings on the large rear esplanade, which is reached by means of steps leading down through the rocky landscape.

Christ Crucified, by Beobide.

The Cross and the Monastery from the Social Studies Centre.

The Rear Esplanade and Buildings

Behind the Risco de la Nava is a vast esplanade of three hundred metres by one hundred and fifty flanked by two rows of thirty-three arches connecting the buildings of the Benedictine Abbey (next to the Risco) and the Social Studies Centre which stands opposite it.

The first Monastery, built by Muguruza between 1942 and 1949, was rejected by the Benedictines on account of the distance between it and the Basilica. Méndez therefore found himself obliged to construct new buildings for the Abbey and the *Escolanía* similar to the original building raised by Muguruza, linking them with long arcades which were based to a certain extent on those that may be seen in the Plaza de las Armas at the Royal Palace of Madrid (which ultimately reveals just how far official classicistic frames of reference had changed over the years).

The Benedictine monks entered the Monastery in accordance with the terms of a contract signed on 29th May 1958 specifying that at least twenty monks should devote themselves to: "maintaining the cult with splendour and to that end direct an *Escolanía* (choir) of forty children, who shall there receive their education; organizing a spiritual retreat; and taking charge of the administration of the Hospice and the management of the Social Studies Centre".

The Benedictine Monastery's first abbot was Justo Pérez de Urbel (1958-1967), who from the very beginning paid special attention to the choir or *Escolanía* as it is known in Spanish and achieved the high standards that have been maintained by its members ever since.

Muguruza's Monastery is an interesting example of the official aesthetics of his time. A comparison of the details of this building with Méndez's (opposite) reveals how the architect's taste was influenced by the Bourbon aesthetics of the Royal Seats.

Muguruza's original Monastery building now houses the Social Studies Centre, which contains a large lecture hall, a library, several rooms to accommodate work groups and seminars, an assembly hall, and a hospice.

The door to the Basilica is a simplified version of the one at the main entrance. Here Egyptian inspiration is evident to a more considerable degree and the side walls take up the buttress motif (although in this case flatter) which was designed for the façade by Muguruza. Behind it a vestibule leads to the extremely long, bare corridors running to the Basilica.

The esplanade, the arcade and the rear Basilica door were built in 1956 by the Huarte construction company.

An aerial view of the Cross and the Monastery with the Social Studies Centre in the foreground.

At the end of the visit to the Valley, it is a good idea to compare the solemnity of the esplanade with the surrounding countryside, albeit on short stops along the roads linking the various points and leading back to the outside world.

BIBLIOGRAPHY

ANDRADA PFEIFFER, Ramón: "Funicular del Valle de los Caídos", *Reales Sitios,* 47 (1976), pp. 65-72.
BONET CORREA, Antonio: "El crepúsculo de los dioses", from BONET CORREA, Antonio (coord.): *Arte del franquismo,* Cátedra, Madrid, 1981, pp. 315-330.
CIRICI, Alexandre: *La estética del franquismo,* Gustavo Gili, Barcelona, 1977, pp. 112-124.
DOMÉNECH, Luis: *Arquitectura de siempre. Los años cuarenta en España*, Tusquets *(Cuadernos ínfimos,* 83), Barcelona, 1978, pp. 46-50, 68 and 72,
JUNQUERA DE VEGA, Paulina, and HERRERO CARRETERO, Concha: *Catálogo de tapices del Patrimonio Nacional.* Vol. I: siglo XVI. Patrimonio Nacional, 1986, pp. 54-62.
MADRAZO, Pedro de: "Tapicería llamada del Apocalipsi...", from *Museo Español de Antigüedades,* Madrid, 1880, pp. 283-419.
MARCHÁN FIZ, Simón: "El Valle de los Caídos como monumento del nacional-catolicismo", *Guadalimar*, no. 19, January 1977.
MÉNDEZ, Diego: *El Valle de los Caídos. Idea, proyecto y construcción*. Fundación de la Santa Cruz del Valle de los Caídos. Madrid, 1982.
MORENO, Juan: "En el Valle del nacional-catolicismo", *Triunfo*, no. 721, November 1976.
Santa Cruz del Valle de los Caídos (Guide). Patrimonio Nacional, Madrid, 1963, 1972, 1983 (various revised and extended editions) and 1985.
SANCHO, José Luis: *La Arquitectura de los Sitios Reales. Catálogo de los Palacios, Jardines y Patronatos Reales del Patrimonio Nacional*, Patrimonio Nacional-Fundación Tabacalera, Madrid, 1996.
SUEIRO, Daniel: *La construcción del Valle de los Caídos.* Sedmay ed., Madrid, 1976.

This book, jointly published by El Patrimonio Nacional and Aldeasa,
was printed in Madrid on 31st July 1998, Feast of St Ignatius Loyola,
at Estudios Gráficos Europeos S.A.

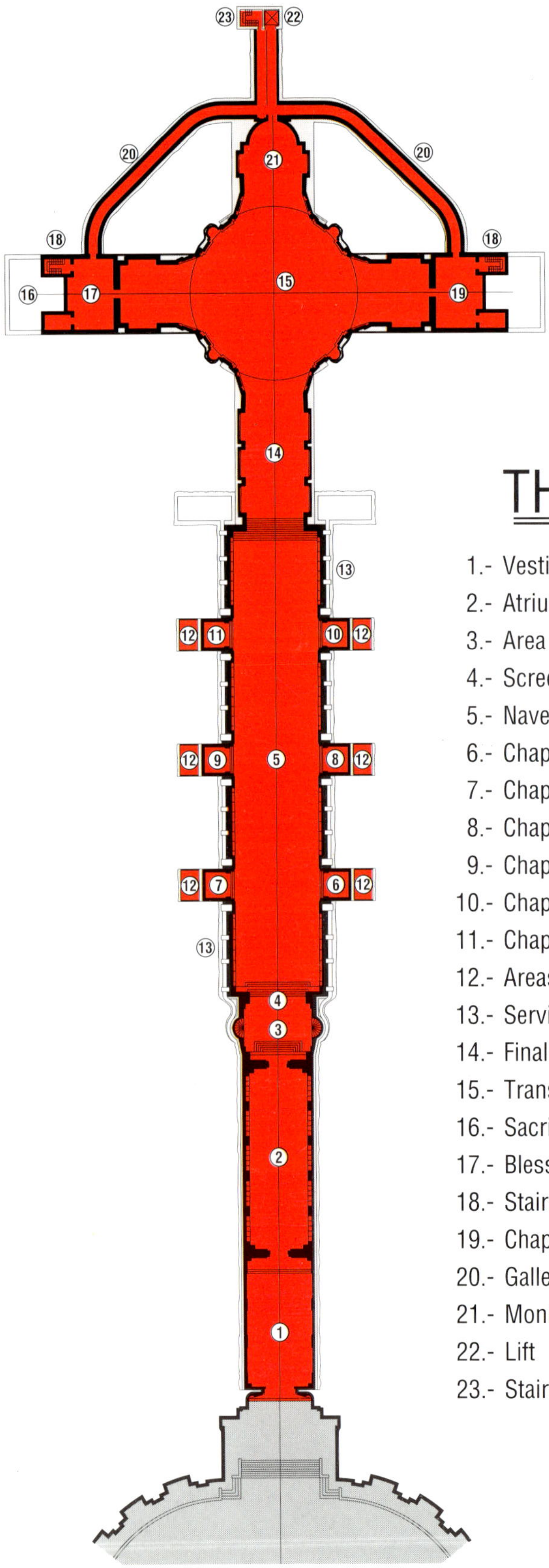

THE CRYPT

1.- Vestibule
2.- Atrium
3.- Area between the Atrium and the Nave
4.- Screen
5.- Nave
6.- Chapel of the Assumption of the Virgin
7.- Chapel of the Annunciation
8.- Chapel of the Marriage of the Virgin
9.- Chapel of the Epiphany
10.- Chapel of the Flight into Egypt
11.- Chapel of the Transit of the Virgin
12.- Areas for future burials
13.- Service Gallery
14.- Final section of the Nave before the Transept
15.- Transept
16.- Sacristy
17.- Blessed Sacrament Chapel
18.- Staircase to the galleries and ossuaries
19.- Chapel of the Entombment
20.- Gallery to the staircase and lift
21.- Monks' Choir
22.- Lift
23.- Staircase